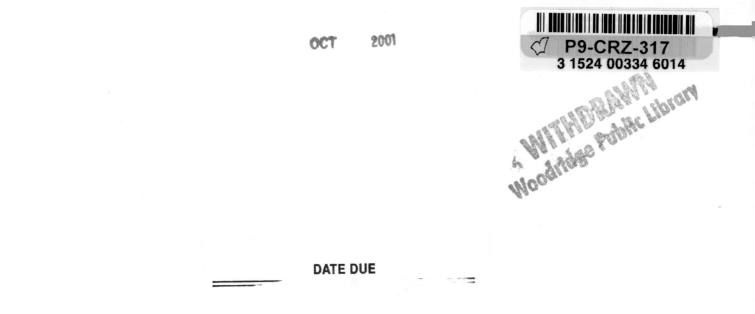

OMNIVORES

in the Food Chain

ALICE B. McGINTY
Photography by DWIGHT KUHN

The Rosen Publishing Group's
PowerKids Press
New York

For my mother, Linda K. Blumenthal — Alice B. McGinty
For Chris — Dwight Kuhn

Published in 2002 by The Rosen Publishing Group, Inc.
29 East 21st Street, New York, NY 10010

First Edition

Book Design: Maria E. Melendez
Project Editor: Emily Raabe
Photographs © Dwight Kuhn

McGinty, Alice B.
Omnivores in the food chain / Alice B. McGinty.
     p. cm. — (The Library of food chains and food webs)
Includes bibliographical references (p.).
ISBN 0–8239–5756–X (lib. bdg.)
1. Omnivores—Food—Juvenile literature. 2. Food chains (Ecology)—Juvenile literature. [1. Omnivores. 2. Food chains (Ecology) 3. Ecology.]
I. Title. II. Series.
  QL756.5 .M39 2002
  591.5'3—dc21
                                                                              00–013025

Manufactured in the United States of America

Contents

Food Chains and Webs

Think of the many kinds of foods you eat. Some foods, such as fruits and vegetables, come from plants. Meat, eggs, and milk come from animals. People eat foods that come from both plants and animals. This means that we are omnivores. Some animals, including sheep, are herbivores. Herbivores eat only plant foods. Other animals, such as tigers, are carnivores. Carnivores eat only animal foods.

If a sheep eats a plant, and a tiger eats the sheep, they have formed a food chain. Whenever one living thing eats another, that makes a link in a food chain. Tigers eat other animals, too. Sheep also are eaten by wolves and coyotes. This makes tigers and sheep part of many food chains. Food chains that are linked together are called food webs.

People are omnivores. This makes us part of many food chains.

Passing the Energy

Energy enters food chains through sunlight. Plants use the energy from sunlight to make their own food. Plants are called **producers**. Animals that eat plants are called herbivores. Herbivores are also called **primary consumers**, because they consume, or eat, producers. Meat-eaters, or carnivores, are **secondary consumers**, because they eat primary consumers to get energy.

Omnivores can be both primary consumers and secondary consumers. For example, pigeons are omnivores that eat seeds and fruit from plants. They also eat primary consumers such as caterpillars.

Every food chain ends when the plant or animal at the end of the chain dies. At this stage, tiny creatures, called **decomposers**, break down the plant or animal's body.

6

PRODUCER

DECOMPOSER

SECONDARY
CONSUMER
(Carnivore)

PRIMARY
CONSUMER
(Herbivore)

Raccoons Are Omnivores

The raccoon is a good example of an omnivore. Raccoons eat vegetables, fruits, nuts, seeds, grass, and berries from plants. They also eat animals, including crayfish, fish, mice, lizards, turtles, tadpoles, birds, grasshoppers, and worms. Raccoons usually hunt at night. Animals that hunt and kill other animals for food are called **predators**. The animals that predators hunt are called **prey**. Raccoons have good night vision and sharp hearing to help them find their prey. Raccoons also have an excellent sense of touch. They reach their paws underneath stones in the shallow water of ponds and streams to feel for crayfish hiding underneath.

Ecologists are scientists who study living things and Earth. Ecologists have learned that animals survive best by eating many kinds of foods.

Omnivores such as raccoons survive well because they eat both plants and animals for food.

The word raccoon comes from a Native American name meaning "he who scratches with his hands" or "the feeler." A raccoon's front paws are very sensitive. Compared to people's hands, the raccoon's front paws have four times more nerves to sense touch.

Adapting to Change

Raccoons have changed their way of life so they can benefit from people. For example, raccoons normally hunt at night. However, they have learned that, at times, it is better to come out in the daytime to find food.

The surroundings in which an animal lives are called its **ecosystem**. Some raccoons live in a forest ecosystem. They make homes in hollow tree trunks near ponds or streams.

Sometimes people cut down forests to build cities. Many forest animals die when their ecosystem is destroyed. However, raccoons are survivors. Raccoons have adapted, or changed, to survive in a new ecosystem. This ecosystem is the city. Instead of living in trees, raccoons have learned to live in attics or under porches. Instead of finding food in grasses and ponds, they find food in garbage cans and vegetable gardens. Raccoons can even open the lids of closed garbage cans. Raccoons find food whenever and wherever it is available. Animals that use every opportunity to find food are called **opportunists**.

Sometimes raccoons make friends with people. The clever raccoons visit whenever their human friends are there to feed them. Raccoons also may steal food from people's gardens and garbage cans. This does not make them a lot of human friends!

Finding Food

Many omnivores survive by eating different kinds of foods during different seasons. In the summer, for example, grizzly bears eat mostly plants. They have big back teeth, called molars, to help them chew plants. Black bears, which are smaller than grizzlies, climb trees to find honey and fruit. In the fall, when berries grow on bushes, the bears eat berries. In the early spring, when berries and fruit are hard to find, the bears eat fish. Bears have pointed front teeth, called canines, to help them catch fish. Sometimes bears catch fish with their sharp claws. Grizzlies also use their claws to dig for roots and to hunt prey, such as baby deer or moose. Bears

During the winter, when there is little food available, bears hibernate. Hibernating is not the same as sleeping. When a bear hibernates, its heart and breathing slow down. If the weather warms up in the middle of winter, a bear may wake up, go out to eat some food, and then return to its den for the rest of the winter.

must spend most of their time eating. In the fall, they eat even more so they can store enough food in their bodies to help them survive the winter.

Bears are at the top of the food chain, which means that no other animals eat bears. They do suffer from their habitats being destroyed by humans, however.

Omnivores in the Water

Ocean, lake, and pond ecosystems have many food chains. There are many omnivores in those food chains. Most of the food chains in water begin with tiny plants and algae called **plankton**. Many omnivores in the ocean eat plankton. Sea sponges filter plankton from the seawater by pumping it through many small holes in their bodies. Schools of small fish eat plankton. **Baleen** whales eat plankton by filtering seawater through comblike baleen in their mouths.

In ponds and lakes, catfish are a common omnivore. Catfish eat plants and animals. Catfish roam pond

Most catfish hunt at night. Their eyes are adapted to help them see in the dark. They also use their barbell-like whiskers to feel for food as they swim.

bottoms. They eat the bodies of dead fish that have sunk to the bottom. Animals that find and eat dead animals are called scavengers. Scavengers help keep the water clean.

Catfish use their rough teeth to scrape algae from the bottom of the pond and to chew up dead or live fish.

Finding a Niche

To survive, every animal in an ecosystem must find food that is not eaten by other animals. In this way, each animal creates a **niche**, or place, for itself. Because omnivores eat so many different foods, they are able to find niches in many ecosystems.

Pigs, for example, use their snouts to root, or dig in the ground for food. Wild pigs crawl into hard-to-reach places to root. This allows them to find foods that other animals can't. Sometimes herds of wild pigs work together to move dead tree trunks in search of snakes, rats, or snails living underneath. Because wild pigs eat many different foods, they can survive in

People have created a niche in our ecosystem by growing our own food. People grow plants for grain, fruits, and vegetables. Farmers raise animals, such as pigs, for meat. Most farmers feed their pigs corn, grains, and soybean meal. Pigs gain weight faster than any other kind of farm animal.

16

many ecosystems. Wild pigs can live in forests, deserts, mountains, and marshes. Pigs on farms root for food, too. They eat almost anything, including snails, worms, insects, and plant roots. In this way, pigs create a special niche for themselves in a food chain.

Mice are omnivores that have adapted to life around people. They search for leftovers and all other available food in the places where people live. Mice in the wild eat seeds, fruits, berries, and insects.

Omnivores That Fly

There are more than 40 different kinds of gulls. Not all of them live near the ocean. Some of them live near farms and eat grasshoppers and other insects that farmers don't like. Gulls can live to be 35 years old.

Many birds are omnivores. For example, robins eat worms and insects, along with seeds and fruits. Diving ducks dive underwater to catch fish and find water plants. In shallow water, shoveler ducks shovel up mud with their wide beaks. Their beaks strain out seeds and tiny water plants while the water drains away. Shoveler ducks also use their beaks to strain out insects and shrimp from the water. Tall flamingos hang their heads upside down and poke their large beaks into the water. Their beaks strain out seeds and algae. Herring gulls eat live fish, dead fish, berries, worms, eggs, and almost anything else. They have been known to take clams from the shore, carry them high in the air, and then drop them on a hard surface. When the shells break, the gulls eat the clams. They even follow ships so they can eat garbage that is thrown overboard. Herring gulls, like raccoons, are opportunists.

The mallard duck uses its beak to strain food from the water. Mallards eat seeds, leaves, insects, and worms from near the surface of the water. Ducks fly south for the winter to places where they can find food.

Omnivorous Insects

Many ants eat a sugary liquid made by tiny insects called aphids. The liquid is called honeydew. The ants stroke the aphids with their feelers, and the aphids release a drop of honeydew. Some ants take care of groups of aphids so they can "milk" them for honeydew anytime they want.

Even insects can be omnivores! Crickets and ants are omnivorous insects. Crickets hunt at night and eat many kinds of plants and insects. Ants eat many kinds of foods, too. Thief ants invade people's homes and search for food. They steal sweet things to eat. Carpenter ants make their homes in wood. They suck sweet juices from plants and other insects. Meadow ants spend much of their time hunting for insects in the soil. Meadow ants work together to kill large insects. When the insects are dead, the ants bite them into pieces. Each ant drags parts of the insects back to the nest. Ants are strong.

Each ant can carry things that weigh up to 50 times more than its own weight!

This house cricket (top) and field cricket (bottom) are both able to find plenty of food in their environments.

There are more than 10,000 kinds of ants. This wood ant is dragging an insect wing back to the nest.

Balance in the Ecosystem

People, like any other omnivore in the food web, affect the ecosystem. If we take care of our forests, oceans, and grasslands, we will help keep the food chains healthy!

Finding food is not always easy. Omnivores survive by eating many different kinds of food. Bears, for example, eat different foods year-round. Wild pigs eat foods that other animals can't find. Raccoons and gulls eat anything they can find! Omnivores play a big role in keeping balance in an ecosystem. By eating other animals, they help to keep those populations from growing too large. By providing carnivores with food, omnivores help to feed other members of the food chain. If there were no omnivores, the food chains would be broken.

Glossary

baleen (buh-LEEN) The structure through which baleen whales strain food.

decomposers (dee-kum-POH-zers) Organisms, such as fungi, that break down the bodies of dead plants and animals.

ecologists (ee-KAH-luh-jists) Scientists who study the way living things are linked with each other and with Earth.

ecosystem (EE-koh-sis-tum) The way plants and animals live in nature and form basic units of the environment.

niche (NICH) The role played by each type of animal in an ecosystem.

opportunists (ah-per-TOO-nihsts) Animals that eat whatever kind of food is available.

plankton (PLANK-ten) Plants and animals that drift with water currents. Many are too small to see without special equipment.

predators (PREH-duh-terz) Animals that kill other animals for food.

prey (PRAY) Animals that are hunted by other animals for food.

primary consumers (PRY-mehr-ee kon-SOO-mers) Members of the food chain that eat plants, making them the first link of consumers in the food chain.

producers (pruh-DOO-serz) Plants and algae that use sunlight to make their own food. Producers are the first link in the food chain.

secondary consumers (SEH-kun-dehr-ee kon-SOO-mers) Members of the food chain that eat plant-eating animals, making them the second link of consumers in the food chain.

23

Index

Web Sites

To learn more about omnivores, check out these Web sites:
    www.aza.org/gallery
    www.nature-net.com/bears/cubden.html

24